Kids Who Are Saving the Planet

by Laurie Calkhoven

illustrated by Monique Dong

Ready-to-Read

Simon Spotlight
New York London Toronto Sydney New Delhi

SIMON SPOTLIGHT
An imprint of Simon & Schuster Children's Publishing Division
1230 Avenue of the Americas, New York, New York 10020
This Simon Spotlight edition March 2020
Text copyright © 2020 by Simon & Schuster, Inc.
Illustrations copyright © 2020 by Monique Dong
All rights reserved, including the right of reproduction in whole or in part in any form.
SIMON SPOTLIGHT, READY-TO-READ, and colophon are registered trademarks of Simon & Schuster, Inc.
For information about special discounts for bulk purchases, please contact Simon & Schuster
Special Sales at 1-866-506-1949 or business@simonandschuster.com.
Manufactured in the United States of America 0120 LAK
2 4 6 8 10 9 7 5 3 1
This book has been cataloged with the Library of Congress.
ISBN 978-1-5344-5647-1 (hc)
ISBN 978-1-5344-5646-4 (pbk)
ISBN 978-1-5344-5648-8 (eBook)

CONTENTS

Introduction

There is no denying Earth is a beautiful planet. But it has many problems. Problems like climate change and chemicals in drinking water.

You might think that you're too young to make a difference. But kids all over the world are finding ways to help fix the damage humans have done to Earth. They're working to save honeybees and birds. They're cleaning up our water. And they're clueing kids and adults in about these important environmental issues. Some of them are even demanding change from their governments.

The kids in this book are just a few of the young people working to make the world a better place. Kids are saving the planet!

Chapter 1
Meet Mikaila Ulmer

You might think getting stung by a bee is bad luck. But not for Mikaila Ulmer. Mikaila was just four years old when she got stung by a bee twice in one week! The bee stings were scary. But instead of giving in to her fear, Mikaila wanted to learn about bees.

She learned that honeybees do much more than make delicious honey.

Honeybees live on the pollen they gather from plants. Pollen sticks to the bees and rubs off on a different part of the plant or on another plant. This is called pollination. It creates seeds for the next generation of plants. Bees aren't the only animals that help with pollination, and pollination can also occur with wind. Without honeybees, we might have to live without apples, cranberries, melons (including Halloween pumpkins!), and broccoli. Blueberries, cherries, and almonds could also disappear.

Mikaila learned that some bee species are **endangered**. (An endangered species is any type of creature or plant that is in danger of disappearing forever.) Scientists don't know why the bees are dying. It might have to do with chemicals used to protect plants from insects that could harm them.

Mikaila wanted to find a way to help the bees.

FLAX SEED

Her parents encouraged her to enter a local children's business competition as well as Lemonade Day in Austin, Texas. Around the same time, Mikaila's great-grandmother sent her an old cookbook with a recipe for lemonade made with flaxseeds. Flaxseeds are believed to be very healthy for you.

Mikaila decided to make her great-grandmother's flaxseed lemonade and sweeten it with local honey for the competition. That's how Me & the Bees Lemonade was born.

Me & the Bees Lemonade was a hit! Mikaila sold it at local fairs. Then a pizza shop asked if they could sell her lemonade. All of a sudden, Mikaila had a business!

From the very beginning, she gave
10 percent of her profits to groups that
help honeybees. Now Mikaila's lemonade
company is called BeeSweet.

Mikaila continued to sell her lemonade around Austin. Then she had a big breakthrough. A grocery store chain wanted to sell her lemonade in their stores!

That same year, when she was just nine years old, Mikaila appeared on the television show *Shark Tank*. She needed **investors**. Investors are people who give money to a company to help it grow, in exchange for a share of the profits. Mikaila needed money to make more lemonade to sell in additional stores. One of the investors gave Mikaila sixty thousand dollars to help Me & the Bees Lemonade reach even more customers.

There were problems as Mikaila developed her business, but as she said in a television interview, you can't let them stop you. "Don't be discouraged by life's little stings, but get back up and spread your wings."

By 2019, at the age of fourteen, Mikaila had already won many awards for young **entrepreneurs**. (An entrepreneur is someone who comes up with an idea and then works to create a business.)

In 2016, Mikaila introduced President
Barack Obama at a conference and was
a guest chef at the White House's Easter
Egg Roll! She was also named one of "The
30 Most Influential Teens of 2017" by *Time*
magazine.

Today, this young CEO (that's chief executive officer) travels around the country teaching other kids how to start their own  businesses, all while she keeps up with her schoolwork!

Mikaila thinks that kids make the best entrepreneurs. As she said in one of her speeches, "Entrepreneurs hold the American dream, and the biggest dreamers are kids. We dream big. We dream about things that don't even exist yet. We believe in our dreams. . . . We believe in the impossible."

Chapter 2
Meet Ridhima Pandey

Did you ever think that the government wasn't doing enough to protect the environment? If you're like Ridhima Pandey from northern India, you would say yes.

But the difference between Ridhima and many other kids is that she decided to actually do something about it.

In 2017, when she was nine years old, Ridhima filed a petition against the government of India.

A **petition** is a formal written request. In her petition, Ridhima said that the Indian government hadn't done enough to protect her and the Indian people from the bad effects of climate change.

The petition was filed in a special court that deals with environmental cases in India. Ridhima asked the Indian government to protect the forests and lower the use of fuels that are bad for the planet.

"My government has failed to take steps to regulate and reduce greenhouse gas emissions, which are causing extreme climate conditions," Ridhima told a newspaper. "This will impact both me and future generations."

Ridhima began to worry about climate change in 2013 when she was five years old. Uttarakhand, the northern Indian state where Ridhima lives, is in the foothills of the Himalayan Mountains. In June 2013, flash floods and landslides in her state killed hundreds of people and left thousands homeless. Ridhima's family and her home were safe, but that didn't stop her from worrying about others.

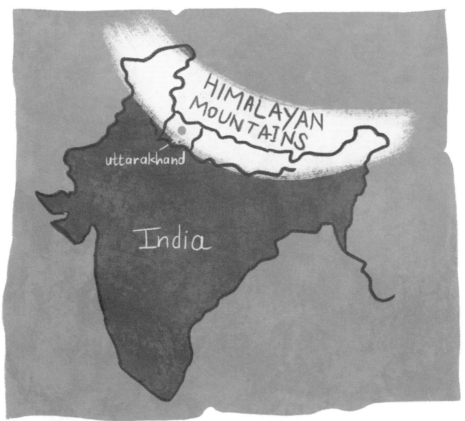

Ridhima learned that the flooding was caused by some of the most powerful **monsoons** in modern history. A monsoon is a period of wind and rain in the Indian Ocean that sweeps over countries in southern Asia. Scientists believe that climate change has led to bigger and more dangerous monsoons. The powerful monsoon in 2013 turned the beautiful foothills of the Himalayan Mountains to mud, which led to frightening mudslides.

And it isn't just monsoons that worry Ridhima. Many cities in India are extremely polluted.

The Indian government has passed laws to protect the country's forests, clean up its rivers, and improve the air people breathe. But the laws are not widely enforced. This upset Ridhima.

"Trees are being cut for roads and buildings," Ridhima told an Indian newspaper. "Industries are polluting the environment; people don't think before wasting water or electricity."

Ridhima thought a petition might get both the government and the Indian people to wake up to the problems they were causing.

In her petition, Ridhima asked that India be forced to follow its own laws. She also wanted to see a plan in place to help correct the damage that had already been done to the environment.

Throughout 2017 and 2018, Ridhima and the Indian government filed opposing claims with India's environmental court. The case was dismissed in 2019, but you can bet that a strong, dedicated young girl like Ridhima isn't going to give up.

Ridhima wants to make the world a better place—for herself, for you, and for future generations.

Chapter 3
Meet Will and Matty Gladstone

Will Gladstone was on vacation in Florida with his family when he heard three woodpeckers in the trees. Those pileated woodpeckers turned him into a bird-watcher. During the next two years, he spotted 190 species of birds, mostly in his home state of Massachusetts.

There's one bird that Will thinks about
a lot. It's one that he's never even seen in
person. It's the blue-footed booby.

The birds are named for their bright blue feet. Male boobies dance around to show off their feet when they're trying to attract females. Females choose males with the bluest feet!

The boobies sleep on land at night but spend their days feeding at sea. They can fly far out to sea looking for small fish like sardines and anchovies. If a booby spots a school of fish, it folds its wings back and dives into the water like a torpedo. Boobies use their long beaks to snap up dinner.

The bird species mostly lives on the coast from northern Mexico to southern Peru. Those birds are alive and well. But there's a small group of blue-footed boobies on the Galápagos Islands. The number of birds in that group has declined by about two thirds since the 1960s. Scientists don't know why.

Galápagos Islands

When Will learned about these birds in his fifth-grade science class, he wanted to help. "They're sort of just a strange and unique bird, and they just have something special to them," Will told the **National Audubon Society**. (The National Audubon Society is a nonprofit organization dedicated to the preservation of birds and the places where they live.) "One day it sort of just hit me that because they have blue feet, you could sell blue socks."

Will's younger brother, Matty, also wanted to help. In 2016, they started the Blue Feet Foundation.

Will and Matty used social media to help sell their bright blue socks—the same color as the birds' feet—with a picture of a blue-footed booby on them. The boys donate all their profits to efforts to help the blue-footed booby. Bird lovers from all fifty states and forty countries have placed orders. They also send the brothers pictures of themselves wearing their socks in places like the Grand Canyon and on the Great Wall of China!

The brothers won the 2017 President's Environmental Youth Award from the Environmental Protection Agency (EPA) and the 2017 John Muir Association Youth Conservation Initiative Award. But what's way more important than awards to Matty and Will is the fact that more people know about the blue-footed booby because of them.

The brothers were fifteen and twelve years old in 2019, and by that time they had raised more than eighty thousand dollars. They helped fund an expedition for a blue-footed booby expert to study the birds in the Galápagos.

"I hope we go out of business," Matty told a reporter, "because that means we saved the blue-footed booby."

But until then, these brothers will be doing whatever they can to help those supercool birds.

Chapter 4
Meet Jaden Anthony

Jaden Anthony from Brooklyn, New York, started to worry about climate change and pollution at a young age. Then he learned about a water crisis in Flint, Michigan.

In 2014, the city of Flint began to use the Flint River as a source of water. In the following year, the EPA tested the water. It turned out that the people of Flint had been drinking water with high levels of lead.

Lead is bad for everyone, but it is especially bad for children. Even small amounts can cause health problems and learning difficulties.

Clean water is something that is very important to Jaden. He was born three months premature in 2005, and weighed less than two pounds. He was also born with a rare form of diabetes. Diabetes is a disease that affects how the body uses blood sugar. Jaden's type of diabetes makes him thirsty all the time. Knowing how much he relies on safe drinking water, Jaden wanted to do something to help the kids in Flint.

"I have the ability to drink clean, fresh water. But not a lot of kids have the same right, or the same luck as me," Jaden told an environmental group called Green for All.

A comic book fan, Jaden decided to write his own comic to draw attention to the need for clean water and to raise money. "*Kid Brooklyn* is a comic book I created to make kids and adults open their eyes about the dangers of the world, like global warming and climate change," he said in an interview.

In the first book in what will be a series, Kid Brooklyn and his friends are given a rare Sun Stone that helps them develop superpowers. They use those powers to keep the planet clean and safe.

Kid Brooklyn's friends in the comics are based on Jaden's real-life friends. Jaden writes the stories with his father, Joseph. They work with illustrators to bring the characters and stories to life.

Jaden was twelve years old when his first book was featured at Comic Con in New

York City in October 2017. Comic Con is a short name for a comic book convention. It is an event where the main focus is on comic books and comic book–related movies and TV shows. Comic book fans attend and get to meet creators and other fans. A dollar from every book Jaden sells is donated to Green for All's campaign to bring clean, fresh water to the families in Flint.

Future issues of the comic will support different environmental causes and education-focused charities. Jaden wants to encourage children all over the world to become scientists and engineers who can fix the problems that come from climate change.

Kid Brooklyn is about the power kids have when they work together to make the world a better place.

"I may not be a superhero. I may not have any magic powers. But I know that average humans can be superpowered," Jaden said. He also has a message for the kids of the world who don't have the same access to clean water that he has.

"Don't get sad. Don't lose any hope. We got you."

BUT WAIT...

THERE'S MORE!

In this book, you've met some incredible kids who are saving the planet. Read on to do some fun activities and learn more about how *you* can help save the planet!

Want to help the honeybees?

Here's how:

1. *Bee* informed! Learn about honeybees and share that info with your friends.

2. Plant a garden with bee-friendly flowers that bloom at different times of the year.

3. Ask your parents to buy organic products instead of products that are grown using pesticides that harm bees.

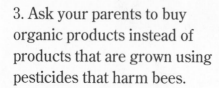

4. Support your local beekeepers by buying local honey. Who knew helping bees could be so delicious?

Is there something in your school or community that you'd like to see changed? Write a petition to gather support.

Here's how:

1. Set a goal. This should be a goal you can achieve, like adding recycling containers to local parks.

2. Write a statement saying why your goal is good for everyone.

3. With a buddy (and with your parents' permission) ask your friends and neighbors to sign your petition.

4. Present your petition to the person who can make the change, like your school's principal or your town's mayor.

You can make a difference for wildlife and the environment too.

Here are some ideas:

1. Educate your friends and family about endangered species.

2. Reduce the amount of plastic you use. Recycle, and buy recycled products.

3. Reduce the amount of water you waste. Take shorter showers. Make sure you turn off the faucet while you're brushing your teeth. These simple changes add up to saving a lot of water!

4. Volunteer your time to protect the wildlife in your area.

Meet another great kid who is saving the planet:
Greta Thunberg!

Greta Thunberg is a seventeen-year-old Swedish climate **activist**. An activist is someone who supports or opposes a difficult issue and pushes for some kind of social change.

Greta first heard about climate change when she was just eight years old. She did not understand why people weren't doing more to stop climate change from happening.

In August 2018, Greta started a "school strike for the climate" outside the Swedish parliament. She wanted her government to take immediate action against climate change.

Greta's strike made headlines around the world. Since then, Greta has not stopped fighting for climate change. She and more than one million schoolchildren worldwide have participated in her school strike movement called Fridays for Future. Greta also speaks at important conferences to address environmental issues, and she has inspired many around the world to take action.

"The one thing we need more than hope is action," she said at a talk in Stockholm, Sweden, in 2018. "The climate crisis has already been solved. We already have all the facts and solutions. All we have to do is to wake up and change."

Greta gives talks all over the world about climate change. Her hope is to inspire world leaders to focus "on the climate itself" and "unite behind the science!"

Now that you've met some of the kids who are saving the planet,

what have you learned?

1. What is the definition of "endangered species"?

a. An endangered species is any type of creature or plant that is in danger of disappearing forever.

b. An endangered species is any type of animal who is not in danger of disappearing forever.

c. An endangered species is any type of animal who is a danger to the public.

2. Why are honeybees important to Earth?

a. Their venom has special properties.

b. They pollinate plants and crops to help new plants grow.

c. They create hives to protect trees.

3. Why did Ridhima Pandey file a petition against the government of India?

a. She thought the school days were too long.

b. She thought the government hadn't done enough to protect Indian people from the bad effects of climate change.

c. She wanted better food in the school cafeteria.

4. Clean air is important to Earth because it helps keep people healthy.

a. true   b. false

5. What is the name of the bird that Will and Matty Gladstone want to protect?

a. blue-footed booby   b. blue-footed baby   c. pileated woodpecker

6. What superpowers do Kid Brooklyn and his friends have?

a. They can fly, and swim underwater without breathing.

b. They can keep the planet clean and safe.

c. They can stop aliens from invading Earth.

7. What is a petition?

a. a speech you give to your parents

b. a formal written request

c. a formal drawn or painted request

8. What are three ways you can help protect honeybees?

a. Learn about honeybees, plant a garden with bee-friendly flowers, and use pesticides on plants.

b. Learn about honeybees, throw honey away, and plant a garden with bee-friendly flowers.

c. Learn about honeybees and share the information, plant a garden with bee-friendly plants and flowers, and buy organic products instead of products that are grown using pesticides.

Answers: 1. a 2. b 3. b 4. a 5. a 6. b 7. b 8. c